ROY G BIV

ASTONISHING POEMS

OF

SCIENCE & SPIRITUALITY

JESSIKA A. SATORI

TWIN RAVEN
PUBLISHING

ROY G BIV

ASTONISHING POEMS
OF
SCIENCE & SPIRITUALITY

TWIN RAVEN PUBLISHING
VASHON ISLAND, WA

JESSIKA SATORI

EBOOK 978-0-9600747-2-3
PAPER 978-0-9600747-8-5

Dedicated to

Fr. Killian McDonnell
and
Sr. Delores Dufner
and
Dr. Irene Hays

Other works by Jessika Satori:

Momisms: Witticisms My Mother Used to Grow My Life

Come to My Dinner Party

Synchronicity: The Entrepreneur's Edge

Synchronicity Secrets

Bumping Up to First Class

I'm a Product of Thomas Jefferson's Seventh Generation

Angle of Repose

Beyond Coincidence

Appellation

So Shall You Reap

INTRODUCTION

ROY G BIV was born out of a radiant constellation—an amalgam of science and the spiritual life buoyed by memory and driven by wonder, curiosity, and astonishment.

Each one of these poems is a colored band gathered from the spectrum of my experience. From wheat fields to submarines, from letters to Carl Sagan to conversations about cardiology, these are field notes from an insatiably inquisitive heart.

This chapbook, structured by the varied disciplines of science, is for those who remember discovering a volcano for the first time, those who whisper to the sky, even those who feel the sacred pulse in a stethoscope, a canyon, or a child's laugh.

Breathe deep. Wonder aloud. May these poems inspire your memories, too.

— Jessika Satori

TABLE OF CONTENTS

Intronaut

Spring kissed the morning in the vast desert of Arizona,
Where I met astronaut Edgar Mitchell,
at a conference about consciousness.
I lamented I could not be an astronaut
hearing the cacophony in elementary school,
"you can't be a pilot
because you are a girl."
I told him I decided to become an intronaut,
She who travels in inner space.

He whispered back to me...

I reside there as well.

I Always Knew That God Lived in the Wheatfields

The Palouse wheat waves over the rolling hills,
looking like the backsides of sheep walking.

As a kid I thought those hills were those of the earth mothers who stretched out on their sides,
hips curved, with Palouse Loess skin.

In Winter, they wear white woolen gowns.
In Spring, variegated green dresses, from lemongrass to deep hunter.
In Summer, a gold so bright causes the eyes to squint.
In Fall, is the deep velvet brown robe and plowed earth of corduroy

This is where my bones are.

Wrinkles

"Breathe deep even if it means you wrinkle." Aimee Nezhukumatahil

Breathe deep even if it means you wrinkle.

Wrinkles are life's rivers that mark your time on earth.

You've lived,

You're living!

Don't waste time with us. I mean, come visit but please, don't dawdle!

Stay the length of a bird's song,

tell us the truth, maybe some lies.

Leave your offering of daisies, dahlias, or daffodils,

smile at a memory we shared,

then go.

Leave this place where the stones mark us.

Live your life!

And--don't worry about wrinkles,

we never do.

Born on Margaret's Birthday

Bored? Sad? Anxious? Lonely?

Investigate the people born on your birthday!

I share my birthday with Beethoven, Noel Coward and Jane Austen,

Margaret Mead too.

I would have loved to have met them all, serve them a luscious banquet, a round tabled dinner so we can see all our faces and hear our words.

I already have another dinner party and invite my personal archetypes, represented by four women and colors:

Martha Stewart Blue Willow Blue, my entrepreneur and pragmatic business woman;

Counselor Troi Mauve, my intuitive scientist;

Vibrant Orange Lady Gaga, the performer and writer, outrageous!

Scholarly Scarlet Margaret Mead.

How do you keep them all inside? people say.

FRYX

Remembering Roald Fryxell

We all cried
the day you died.

All the townspeople
the professors on the hill

the farmers and ranchers
the bankers and shopkeepers.
Typically stoic, dried-eyed people.

I was a fourth grader when it happened.
Your wife was the fifth-grade teacher
in the adjoining classroom.

We wondered why the teachers were huddled in the hallway,
why classes started late.
They corralled us in the library,
that's when we were told.

We cried in the classrooms and the halls
the lunchroom and the bathroom stalls
where we wiped bitter tears on the sleeves
of our starched white cotton shirts and blouses.

Even though we may not have known you personally,
our cheeks puffed, our chests swelled,
you made us proud.
We watched you dig at the Marmes Rockshelter
knew you studied moon rocks and more,
the vast work within in your short life.

Now you are honored by a crater on the moon named after you.

You belong not just to us,
you belong to the universe.

Billions and Billions: I Wrote a Letter to Carl Sagan, and He Wrote Back

When I heard of his diagnosis,
I was determined to help.

I read his book Cosmos over and over.
a teenager working at Frederick & Nelson,
I asked the bookstore manager to hold a coveted copy to me,
to gift to my mother for Christmas.
Even though the chances were one in a billion
to get a book,
the manager got an unexpected shipment,
hid it in the storage room for me.

A national, international and galactic treasure,
I had to help in some small way.

I wrote you a letter.
I offered my home to you and your family,
on a bucolic island
away from prying eyes, just to rest.

A short time later,
I opened the mailbox.
A letter from Cornell University.
I was so excited that the sun and the Douglas Firs
changed in the bright light,
even they wanted to know what the contents held.

I sprinted up the long, gravel driveway,
gravel spitting from under the soles of my shoes,
this precious missive in my grasp,
flung open the screen door, letting it slam,
I sat down at the dining table,

took a deep breath,
and opened the envelope carefully.

A one-page letterhead
thanking me for my kindness
to offer such a place. I read the words over and over,
traced my finger over your signature.

Even though you are with the stars.
I still have your letter among my treasures.

Jefferson Elementary

Everyone lit up when she mentioned the word "candy."
Mrs. Wriggle was my first-grade teacher,
but first of all she was an artist, so she painted the classroom loud.

We made candles, kites, but the best project were the piñatas we made in classroom six. It was for the science project about the solar system.

The discordant bell rang to start school.

She gathered us around a cauldron of porridge-like goop,
and whispered of the magical properties of papier-mâché.
I remember the chill of the goo as I slid the excess off the strips of the Spokesman-Review, placing slimy bandages on the balloon layered with the music of laughter that hung over our heads.

That afternoon we did our spelling tests under the orbs.

The last hour they were dry enough to paint.

Flinging the door open the next morning, our eyes darted around the room our mouths, open, at first silent, then to wailing and moaning -- the result -- many of the planets collapsed.

Neptune was a blue raisin shriveled next to Jupiter. Jupiter hung with its eye cast down.Pluto, an isolated bowling ball, curved in on itself.

The Sun, the planet I was so happy to work on, was a massive half-dried peach hung by a noose that stretched almost to the teacher's desk.

Mrs. Wriggle told us to put out things down, dry our eyes, gathered us around, then we walked to each planet, and she talked about each one.

To salve our disappointment, she filled Saturn with candy during lunch.
we moved the usually meticulously rowed, staid desks back from the center of the room,

blindfolded the tallest boy, the most handsome and swoonable David Star,
and when he hit ribboned piñata with a broomstick, it exploded with a loud crack

all twenty-two of us scrambled without mercy, to score a Bit-O-Honey or Jolly Rancher, our mercenary nails scratched each other's skin, even to blood letting.

Greedily grasping our spoils, we stood by our desks and smacked and suckled the candy until the bus arrived for us to go home.

Rx

Hyperbaric Chamber = six hours a day.
Femur'd gashes carve canyons
that can swallow my finger.
She once encircled her babied fingers
around mine.
Then heroin ensnared her.

Flesh-eating is what they call it.
I call it the eventual taking away
of the sweet-smelling powdered skin
Entirely dependent on me.

Again, I make all the decisions,
Thick with interminable fear.

Her treatment will be forever
if she lives;
My grief will be forever if she dies.

Snap-on Body Parts

How I long for the day
when, *finally*
there could be DIY Snap-On-Body Parts
like Snap-On-Tools
and have
that can be delivered by van
to our doorstep, instructions included.

Replace that knee?
certainly.
Replace that ear?
so I can hear.
A crushed shoulder?
no longer a boulder.

In our dreams
and in certain labs
these probably already exist.

I await distribution.

G. Grew the Sweetest Carrots

G. grew the sweetest carrots
O. grew the brightest peonies
L. grew the largest dinner-plate dahlias
D. grew the most bodacious tomatoes.

All of them had their gifts and they *lent me their bow.*

My Father's Sacred Heart

At 7am, He howled, clutched his chest,
fell to the concrete floor of the machine shop.

The hired man came early that day.
He happened to be in the shop and not outside.

My brothers curiously were there
they usually left to be in the field by now.

The medic happened to be working a half mile away.
He raced over to help.

The ambulance came and whisked him to ER of a local hospital.
The doctor had sense and humbly recognized it was out of his league.

The doc called the Airlift.
Coming from Spokane would be too late.

It happened that another Airlift was in Lewiston.
Loaded him up, whirled him to Spokane.

The medical centers name?
Sacred Heart Medical Center. What else?

Surgeons happened to be there and
slowed his heart to a less than a crawl.

Sewed his burst aorta closed.
A procedure. A miracle.

I happened to be in the Santa Barbara airport at the time.
An interior voice told me not to get on the plane.

Because I didn't, I got the important phone call.
And I rerouted through LAX.

After surgery and rehab,
Dad went back to the fields.

In the words of Raymond Carver, "it was all gravy"
We had him for two and a half years longer.

When discouraged people ask me
Where is God in all this trouble on earth? Where is the proof?

I just point to my Dad, 85 years old, who survived.

The Chemistry Set

One thing I will never forget

was Susie and me and the chemistry set.

We were in science class on the top floor of the Jr. High

So many floors it was part of the sky.

Our teacher told us we had time to create an experiment

little did we know it could have been to our detriment.

We ignored the red and yellow warning signs

Shook all the vials to make a design

Decided to create our own potion

Silly with giggles we went through the motions.

Mixed all the chemicals together

to see what kind of substance it would measure.

The teacher came over soon after

Eyes wide thinking how to avert this disaster

He'd have to act quick!

The timebomb was sick!

Ran it to the next room, threw the concoction away.

And the class lived to see another day.

Dinner on the Eiffel Tower

His arrogance fully stuffed in his suede shirt,

He pranced around the halls of the hostel in Paris,

trying to find someone, anyone, to sew his ripped jacket.

He found me: "I hear you know how to sew."

"I'll pay you to do stitch it up. How much?"

"Nah, I don't want pay."

"I want dinner on the Eiffel Tower."

His eyes widened into saucers. Then shoulders softened to a shrug. "OK."

I handstitched up both long sleeves, it took over an hour.

That night we went to dinner on the Eiffel tower.

Trickster

When will things trickle up instead of down?

says the jester wearing the crown.

Sequoia and the Selkie

He was a tall tree.

I loved him tenderly.

Impossible.
I, a selkie, a seal woman,
swathed in the leather of the sea.
I knew my skin could be sold
I knew I could be "rescued."
but the wild-ness and wilderness could never be taken from me.

One day it happened.
I felt the wind and a hand of a man
when my skin was taken from me.

I did not see what happened to my leather skin of the sea

And we were happy
but something went wrong
but I could not hear the water
or sing my origin song.

Time passed always knowing something was missing.

I met a friend while walking in the woods
he was taller than the sky
massive, his skin hardened like a shell.

I learned I was taken from the sea
and into the woods where the giant trees live.
And I was him and he was me.

I sat in his shade of the trunk
plying him with tales from the sea.
And I was him and he was me.

He could see the sea, he said
from the mountains.

Come up! Look!

But I was not used to being so high
he said that the rain brushed his teeth
and I said the bark kept me warm.

We were so different yet we got along
and taught each other our origin songs.

Then one day I was taken away
no way to say goodbye, no way to run
I was at the end of a gun
My tall tree friend.

I hear you calling me
I know you can hear me.
I know your smell,
I know you
I know.
I
I long for you.

Then, one day I found it in the back of a shack
it was my skin,
I'm going back! Going to the winds and the waves
no longer being a slave.

Now I'm back in the waves.
but my love for your is still great
Even though we have different forms
I still feel your bark to keep me warm.

Hells Canyon
"I took you here because you will not see this again." Dad.

We piled like little link sausages into the station wagon.

We just had eaten link sausages for breakfast.

We were going to go on a little vacation, something we did not do very often.

We were going to go to Hells Canyon,

The roll of the river that separates Oregon from Idaho,

and a slice of Washington State.

We would to see it before the dam would render parts of it underwater.

We drove there through the town of Anatone and down Rattle Snake grade.

The name says it all, full of switchbacks that sets off an inner alarm,
a frightened scream while going down down down.

Or was that my brother's muffled scream trying to scare me?

Anyway, we went there to see the how low and long the canyon was.

Instead of scared, I was delighted to see the jagged-toothy mountains,

Named He Devil and She Devil. In my six-year-old mind I imagined the devils lived
on their respective mountains holding their pitchforks and watching the humans in their cars pass.

Soon we drove up to the He Devil Motel. I wondered, must I confess to the priest that I stayed there? Next time when I'm in that dark wooden box at church? Would my parents get into trouble if I said anything?

The potent smell of propane hit like a firehose as we opened the door. Mom opened the windows first thing before making sandwiches.

We slowly drove the length of the river. "I took you here because you will not see this again."
Dad made me realize that things change and savor them while they are here.

Kata Juta
Australia, 2002.

Red rocks so pungent with color

Sky so sapphire-saturated with wonder

Stark! the whoop whoop whoop

A delirious dervish boomerang making a loop

of a being who I have no right yet to know.

The Day Alan Ran Away

Adams School, red brick, two stories tall, the kindergarten upstairs.

We loved the big room dark stained wood with a white circle painted on the floor. Mrs. Batt led us in there one morning and said we were going to play a game. We went into the playroom but something felt different that day. The chairs in the circle. We all sat down looking at each other. Just then two nurses burst through the double doors, starched and serious, carrying boxes.

Traitors! We were duped.

Alan knew what was up.
He did what we all wanted to do.
He sprang out of that rickety folding chair, like a gazelle,
Turned and sprinted like the wind around the back of the chairs.
To this day I never knew why he didn't run for the door,
but he didn't.
One of the nurses put down the needle and ran after him.
"Go Alan Go!" we shrieked!
"Run Alan Run!"
Some of us jumped up and cheered.
Then his eyeglasses betrayed him. His glasses betrayed him!
They slipped and he hesitated
The other nurse ran from the opposite direction.
The first nurse grabbed Alan around the waist,
his moccasins flopping off his feet.
She sat in a chair and kept in him in a death grip,
his striped sweater twisted like a barber pole.
The second nurse caught up and jabbed him
and he let out a howl that curdled the milk in the next town.

Just then,
Mrs. Batt flew in
"What is the racket!"

"Oh nothing" we all said in unison, including the nurses.
I never trusted Mrs. Batt after that.

Box of Notches

Do you know what notches are, the apparel kind of notches?

The ones where it shows on a pattern where to join a sleeve to the armhole,
the side seams together, the collar to the neck?

I lived and breathed sewing growing up,
went to college and majored in design.

I went to my first job with benefits
at an apparel factory, hidden in the toothy grin
of the Wasatch Mountains of Utah.

It was very different, huge warehouses with a flurry of
activity inside, the worker bees with titles such as
cutters and patternmakers,
industrial engineers and sales people.
Other buildings held bombastic bolts of fabric,
fleece and t-shirt jersey,
polyester double knit,
ripstop and Goretex.

On the first day of work, unsuspecting people
were asked to get to go to the cutting room to get a box of notches.

Everyone is in on it, this secret:
There is no such thing as a box of notches!
But the new employee didn't suspect
that it was all a ruse.

Sent to the plant manager first,
"Can you tell me where the boxes of notches are?"
"No, go ask...."
who pointed one of the cutters.

The cutter pointed to the marker maker, who pointed the sample maker,
who pointed at the fork lift driver.
this could go on easily for at least an hour.

An initiation, of course, and for a while after they
found out, and some were red-faced upset.

Welcome to the Company.

Times God Saved my Life: Motor Vehicles Edition:

1. FIAT 500 POP: Oncoming tow truck lug nuts failed, the axel and one of the wheels ended up floating in the Columbia River, the other sheared off the passenger side of my car.

2. TOYOTA PICKUP: One morning the U joint of the driveline broke in the driveway and not going 70 mph on I-5. The mechanic dryly said, "It's a good thing you believe in God."

3. SAME TOYOTA PICKUP: Eastbound I-90, Zooming down Vantage Hill. The brakes failed. Tried shifting down to slow the pickup. It was 11pm and a weekday night so small amount of traffic. Miracle.

4. MIATA:L.A. I-405 the tire blew with two tractor trailers on each side then trying to pull over to the shoulder and I had to wait for a nut job in the right lane going at least 100mph to pass.

5. SAME MIATA: Going northbound over the Grapevine by LA. Noticed there was no Southbound traffic, then over a hill saw the accident and all of the first responders in my lane. Slammed on the brakes to avoid rear ending all of the flashing red lights.

6. SUBARU SEDAN on the I-5 ship canal bridge in Seattle when we were all going 80pm on a Saturday afternoon. Someone stopped ahead of me. Slamming on the brakes, no cars touched mine.

I Remember Bill Rember

Bill Rember,
Tall as a Doug Fir,
a smile like a Clark Gable.
He found clues within the earth
the Clarkia place
where treasures
beckoned him.

Emerald Creek was close by as was
Garnet Gulch.

Now, as I hold the golf ball sized garnet in my hand
gathered from the North Fork of the Salmon River,
I am in thanksgiving for teachers
that not only taught their passion,
but more often than not,
taught about life.

We Had a Talk About Your Brain

"We had a talk about your brain" my students said the second day of class.

I replied, "Anything you care to share?"

"It's the weirdest one of any professors I've ever had."

"I'll take that as a complement."

My brain twists the words

Calgary, Calvary, Cavalry

Like a washing machine spin cycle,

what keeps them separated--the whites, colors, and brights

is that I say these words out loud. Repeatedly.

When I write a sentence like "I want to finish that jigsaw puzzle tonight"

The zz's sneak up and words on the page say

"I want to fizznish that jigsaw puzzle tonight."

That is the way it has always been.

I worked through the pain and shame

that other people layered on it.

That is where the agitation cycle came from.

"Why can't you be like other people?"

Because I can't.

Nor do I want to be.

Under the Sea

We binge watched all the submarine movies we could to prepare for our pilgrimage.

Hunt for Red October
20,000 Leagues Under the Sea
The Abyss
and the Daddy of them all, *Das Boot.*

That was the night before. I woke giddy the next day and left my car in the park and ride while my co-worker's chariot, a PT Cruiser, drove to the submarine dock. I thought we'd only have an hour of someone walking us through a museum, make a donation, then go home.

But they said, yes, we've been expecting you. This way. Ominous.

They took us out to see the nuclear subs lying side-by-side like cigars in a box in the Puget Sound. We followed, walking and walking.
Then they motioned us toward the jaywalk. They are letting us in!

Clouded in claustrophobia, repeating my mantra of coping words, I turn to face the innards of the beasts and shimmy down.

It doesn't take long to realize we are inside a floating city,
I skim my hips on the tiny tunnel of hallways,
stepping over the oval entry from one area to the next.

The tour guide takes us through the galley,
the sleeping rooms where soldiers, like layers of sardines, sleep.

The tour seems like minutes instead of hours.
Up the stairs into the fresh air again.

How do they do it for months at a time?

Kids in the Cart

I was drawn to their chubby cheeks, their puffy thighs and
impossible flexibility.
The kids in the cart in front of me at the tiny grocery.
I saw the jugs of milk, orange juice and loaf of bread hugging them.
Then I saw the horror in the mom's eyes and the panic in the cashier's face.

The State did not pass the budget, the fiscal year hung in the fulcrum,
Citizens swaying on the balance. The mom's food card did not work.

The kids did not cry, but eyes looked at their mom, tears running down her cheeks.

Then light then became so bright, brighter than the white blaze around the meat counter.

Loud like the red neon "OPEN" sign.

I put two and two together. The kids were not going to eat today.

The obvious tale.

The cashier looked for some way out.

It came out of my throat before I could think:

"I'll pay for it."

The expressions switched. The Mom was horrified, the cashier relieved.

"Oh no, no, no." The mom said.

"Yes, please, you need the groceries." I replied

"Oh no no no. It's OK."

"No it's not. Let me. Please let me."

A long pause.

Then.

"I have been where you are now." I said without flinching.

"OK. I mean, I don't know you, I can't."

I reasoned, "you just pass it on to someone else in the future."

That was the master key that opened the gate.

She breathed and looked at those tender cherubs."Yes."

Relieved I gave her a way out of her shame, the hurt of having to take a "handout" from a stranger. But,

I cared about her. I cared about her kids. *I cared about them.*

He Never Shakes

He never shakes
his faith stakes him to the ground
where the earth spills tomatoes and corn,
the cedars and Doug fir cast shadows,
and the shy maples drop their golden gloves vying for a suitor.

He never shakes
his faith stakes him to the saints
who draw pictures of him with carpentered knuckles,
wrapped in a pomegranate-beaded rosary.
He protects what he has,
and what he does,
and what he sees in the dappled light through the apple trees.

His faith never shakes
it makes shelves, promises,
light of a subject,
and when he hears the soul's whistle,
he stops to walk his Camino trail.

His faith never shakes
he awakens our rude, boastful pride and
limping life stories
to the mysteries
of the tomato--
a vibrant red balloon
in the hand, in the salad, in the belly.

He never shakes
the earth quakes because life is now uneasy, unsteady.
When you've gone to as many treatments as he has,
you've earned the angels holding you close,

the sun to warm the weary bones,
and the rain, fragrant with roses, growing sweet gardens of sound.

He never shakes
his faith stakes him to the ground.

ROY G BIV

How on Earth and in Heaven
did Isaac Newton react,
when,
for the first time,
he saw the sun
split into color in the palm of his hand,
spilling to the floor,
spreading a transformative act that lit the room?

What on Neptune and on Venus
tarrying with thoughts being
a mere mortal creating color,
did Newton feel
when he deftly shoved that bodkin
between his eye and eye socket to
"induce the sensation of color?"
All I can think of is the sensation of pain.

What on Saturn and on Jupiter
was he reading in his 30 bibles,
trying to dissect the voice of God?
What colorful words leapt out to give him clues?
How can I decipher the rich resonance that drew him in?

What on Mercury and on Mars
was Newton thinking when he wrote of action having
an equal and opposite reaction?
Pressing on a stone with your finger, the stone presses back.
Thus, as the prism is pressing on me as I hold it,
am I impressing my secrets on it?

Barefoot on the grass,
gazing at the starlit universe on a Palm Sunday dawn,
I mull over of what Isaac wrote
and consider what God created.

The Specters

The Palouse the wheat waves to me,
It was taller than me, and inviting me into its refuge,
I could stretch my arm and my fingers skim the top of the beards,
Encircling and embracing me.

It was Ireland-green in the Spring,

I spoke with God there. I thought everyone did.

Other times, in the late fall, deep fog would settle in.

One of those days my father came in from the fields for lunch, face caked with dirt, his fingers round like sausages.

He told me an astonishing story...

It was early in the morning on that autumn day, he was on the tractor, half looking forward but mostly looking back, just keeping the lines of the plow straight with the last before round, butting just so there were no skips, no overlaps.

Turning forward, a gasp came from his lips.Squinting, he slowed to see something in front of him, actually many somethings in front of him, unreal, not of this world.

An apparition.

Like a scraggy gaggle of specters, it was a group of sandhill cranes staring at him through the fog.

He slowed the chatter of the orange Allis-Chalmers crawler tractor to a stop.

He didn't move and they didn't either. The only sound was the diesel engine, the smell of the exhaust next to the front window, the t-shirt, flannel shirt and jeans wrapped around him.

For several moments they exchanged breath.

And when the sun cracked through the mist they flew off.

Dad told me that it was at the place
where several of the neighbor's ashes were spread,
in the land they farmed and loved.

And in a few months, he was gone to where they went.

I think of that story more often than I can tell you.

The Only Thing Between Life and Death is Breath

A riddled poem

I choke on cigar smoke
walking the squeaky corridor
of a somber Victorian mansion.
When I approach the end of the hall, the French doors open for me, and I see:

On the right, Goethe of the "More Light!"
and Roethke of the "Was It Light?"
converse about verse with Jim of the Lighted Fire listening and nodding next to the fireplace.

At the table, Jesus Light of the World
ladles soup for Madame Curie of the Radium.
She takes the bowl and offers in to Apollo, who beams.

The Brothers Lumiere and Monsieur Daguerre, discussing fascinating images,
their forks full of Crepes Suzette, dancing, punctuating their words;

On the left, Benjamin of the Key and Queen Noor of the Arabic Light take
coffee at the bookshelf while Hank Who Saw the Light
surveys the crowd while he noodles and hums. He tips his hat to

Paul of Tarsus who discusses the finer points a flash of light and horse behavior
with Faraday of the Electromagnetic Force; seated next in the corner is Sir
Isaac who reaches to offer a prism'd dessert to Sir Herschel of the Infrared.

I note how Edison and Tesla stand at opposite ends of the room, backs to each
other, yet an occasional lapse to look at each other triggers daggered lightning.

In the back of the room Eagle-from-the-Light perches alone above the din,
watching the antics.

As I pass through,
none of these souls notice
who I am, an inconspicuous servant igniting
the multitude of candles.

I, Jessika of the Sudden Light,
will be the only one to blow them out.

The Only Thing Between Life and Death is Breath, Explained

I choke on cigar smoke
walking the squeaky corridor
of a somber Victorian mansion.
When I approach the end of the hall, the French doors open for me, and I see:

From the right, Goethe of the "More Light!" [Wolfgang Goethe's last words were "More Light!"]and Roethke of the "Was It Light?" [Title of one of Theodore Roethke's poems]

converse about verse with Jim of the Lighted Fire listening and nodding next to the fireplace. [Jim Morrison of the Doors, known for the song, *Light My Fire*]

At the table, Jesus Light of the World
ladles soup for Madame Curie of the Radium.
She takes the bowl and offers in to Apollo, who beams. [Apollo is the Greek god of Light]

The Brothers Lumiere and Monsieur Daguerre, discussing fascinating images, their forks full of crepes suzette dancing to punctate their words;

[Brothers Lumiere, inventors of the cinema camera, Daguerre,
the inventor of the Daguerreotype, a form of photography]

To the left, Benjamin of the Key [Benjamin Franklin]
and Queen Noor of the Arabic Light [Queen Noor of Jordan. Noor is the word for light in the Arabic language]
take coffee at the bookshelf
while Hank Who Saw the Light [Hank Williams]
surveys the crowd while he noodles and hums. He tips his hat to

Paul of Tarsus who discusses the finer points of a flash of light and horse behavior
with Faraday of the Electromagnetic Force; seated next in the corner is Sir Isaac who reaches to offer a prism'd dessert to Sir Herschel of the Infrared.

[Michael Faraday] [Sir Isaac Newton] [Sir William Herschel]

I note how Edison and Tesla [Thomas and Nicola]
stand at opposite ends of the room,
backs to each other, yet an occasional lapse to look at each other triggers
daggered lightning.

In the back of the room Eagle-from-the-Light perches alone above the din,
watching the antics. [Warrior of the Nez Perce Native American tribe]

As I pass through,
none of these souls notice
who I am, an inconspicuous servant igniting the multitude of candles.

I, Jessika of the Sudden Light,
will be the only one to blow them out. [Satori means sudden enlightenment]

Coronation

Hats! Hats! Hats!

Flock into Westminster Abbey.
Saucer'd, flowered, netted,
in hues of pink and periwinkle.

The bugles bleat:
"We are here to crown a King."

The procession entered, wrapped in capes of
royal blue and cloaks of scarlet velvet,
the shoulders circled in ermine.
The leader of the House of Commons,
a modern-day Boudica.
Her hairpiece, redolent of
Caesar's laurel of victory,
bears the multi-colored jeweled
Sword of the State.

The clerics file in,
trimmed in traditional garb.
Yet there is something different.
it is the red-haired bob of the
female Anglican Bishop of London
who proclaims the Gospel.

The rich sonorous voice of the Welshman,
topped up by the Holy Spirit,
delivering the Kyrie Eleison in Welsh;
The Gospel singers,
this bevy of white swathed doves,
praising Alleluia;
the African yellow butterfly,

singing the sacred fire,
lifted above the gathering.

If we squint,
we can see other visitors from the next world.
Mahatma Gandhi peering over the shoulder
of the Indian and Hindu Prime Minister,
reading the first scripture.

Queen Elizabeth
in proper suit, pillbox, and purse
beams proudly beside the new king.

Diana, black velvet dress and pearl choker,
leans on one of the tombs, watching like a hawk
over her sons and grandchildren.

This could have been her moment.

Other moments carried in memory:
One of the sons, after pledging
to be a "liege of life and limb"
on his knees, arose and swooped in
to plant, on Father's cheek, a kiss;

Clerics, including a Greek Orthodox Priest
and a Roman Catholic Cardinal gave blessing!

A young lad stood transfixed,
recalling David and Goliath,
face to face with the new King:
"Your majesty, as children of the Kingdom of God,
welcome you in the name of the King of Kings."

A few heart beats,
and it was all over.
A grand and gallant gesture placed in a sacred envelope,
forever sealed in space and time.

And as they recessed, the bells pealed
and the Savior's vow,
"I come not to be served but to serve"
rang too.

Out in the rain, the twitching flanks
of the Windsor Greys appear,
spelling out a staccato'ed code:

Hooves! Hooves! Hooves!

Confession

The howling wind shouts out my illusions...
disordered attachments,
blatant neon'd failures,
ongoing near and far occasions of sin,
forfeiture of common sense concerning the custody of the eyes,
flagrant disobedience of the basic tenets of Christianity.
Outrageous tone and volume while driving in traffic.

I am guilty of all of these.

Oh...yes...and...

Promotion of sleek, steel-toed lies.
Attempts and successes at stoking the brazen fires of discord.
Lording over mine and other's schedules with incessant bleating.
Souring countenance at any novel idea.
Crying to induce manipulation
Pouting at the least corrective phrase.

I am sorry for these and all my sins that I wish never to expose.

The Scarab

I built my thoughts on Jung's thoughts
first reading his Memories, Dreams, Reflections
wrapped in the plastic cover
from the public library.
This is where I first read about the bookcase sharply cracking,
the scarab buzzing, tapping a morse code on the window,
and the synchronicity of the many fish visits in
a short span of time.

Later, I would create my doctoral work looking
at the role of synchronicity, how influences
entrepreneurs, and I would fill journals
full of examples from my everyday life.

The scarab's blood courses in mine.

Dinosaurs

"Brontosaurus, Stegosaurus, Tyrannosaurus Rex
Which is the one we like the best?"
--3rd grade Blue Ladies Chant

Remember the player piano?

The paper rolls, perforated, carefully unwound and threaded
above the keys on the piano. Dot dot dot.
As kids, our astonished eyes watching how the piano was played without
a human!
How does it do it? we'd say to the grown ups, looking up to them.
How did the piece of paper play and keys and make the sound?

Later, when my mom was studying for her masters degree,
she brought home a mystery.

It was shaped like a hand-held tape recorder, maybe 5x10,
and instead of cassettes
there were rolls of paper, like
a small player piano.

And I threaded the paper through the gears,
remembering how the arthritic hands of the elders
wound it on the piano,

but it was different.
There were buttons on the side, ABCD,

and the paper wound to the first picture
and asked, what dinosaur is this?

A) Brontosaurus
B) Stegosaurus
C) Tyrannosaurus Rex

and you would press the button,
and it would to the answer screen.

Still now, I wonder who did invented this,
a prototype for a personal computer.

If Wishes

I wished with all my heart heart that I had a horse horse. I know this is so because my family doctor gave me a book: Black Black Beauty Beauty, for being brave when I got my tonsils out out. I was six just just six. I loved these strong strong massive 4 legged-souls. I wanted to ride. We didn't have any at the farm I grew up on. I wanted one so so so so. But we were crop people not ranch people. But still still still I dreamed. A horse with a muzzle nuzzle puzzle. Snout out snort nostrils flame hot like fire fire engines when they gallop gallop. I wanted a horse so bad bad. Posters on my wall. Drawings I sketched. I'd would have given up a month of hot fudge sundaes just to have a ride. Just just one ride. I coveted what my friends had. The resin'd stallions stallions and mares with chains chains for rein reins. One so called"friend" had over twenty that took up up a whole wall in her bedroom. She contorted snorted like she she was better than me me. Rolling eyes eyes and clicking, bray braying.

As I grew older. I wanted my first first very first car car to be a Ford Mustang. What did you expect? The one where the silver 'stang emblem shone like a silver silver sliver in the sun sun. The unique back turn signal three coordinated lights that shouted it's own morse code: Dit-dit-dit. Dit-dit-dit. Later I went to the State Fair plenty of cacophony I found there. Plenty of cacophony cacophony and cotton cotton candy-cacophony candy. Bursting through the entrance gate, I made a beeline to the horse barns. Where I saw the loco loco motive-like flanks of the Percheron, the shaggy-haired hooves hooves of the Clydesdales, the dotted Appaloosas looses Appaloosas and the silk-silky gnarl-gnarly sleek Freisians. Horses. Storms storms on four legs—lighting snap and thunder clap. Matched the intensity of my sacred scarred scared soul. The harshness of teenage angst. Buried in the mane of a dream.

If I should die before I wake, my horse will heaven-ride make.

The unbridled biography of Jessika Satori

Dr. Jessika Satori is a multidimensional force.
A poet, writer, artist, educator, and adventurer.

She was raised in a sea of golden wheat
then migrated west
to be close to the cyan sea.

Jessika studied storytelling with:
screenwriter Stewart Stern,
poet Anne Waldman,
and songwriters Holly Knight, and Sharon Vaughn,
voice with Danny O'Keefe and Holly Boaz.

She studied book binding with
Sandra Lopez, abstract painting with Peggy Zehring
calligraphy, Japanese monk Kazuaki Tanahashi and
artist and educator Victor Moore.

Jessika created "The Poetry Project" when working as a
chaplain in Harborview Medical Center in Seattle, WA, and
is been a retreat director, especially in the areas of Ignatian Spirituality,
sacred art and Celtic Spirituality.

Dr. Satori wrote Synchronicity: The Entrepreneur's Edge, a business
decision making book with Butterworth Heinemann Press and
Momisms: Wittisms that My Mom Used to Grow My Life, a memoir about
her mother, and produced four poetry chapbooks:
I'm a Product of Thomas Jefferson's Seventh Generation,
Angle of Repose: Poems of Geology,
Come to My Dinner Party, and
Everything is Written on the Sands of Heaven.

In her other lives, Dr. Satori has been a college professor
of business in the USA, UAE, South Korea,
Russia, and Madagascar.
She has spoken and read poetry in venues worldwide.

Her award-winning films, where she has been in the roles of
producer, director and writer,
have been shown as a part of festivals in Ojai,
Yosemite, and Pasadena CA and released
worldwide.

Jessika is an enthusiastic traveler,
reaching all seven continents and fifty states.
She traveled to Easter Island for Easter,
Christmas Island for Christmas,
Ascension Island for Ascension,
and Pentecost Island for Pentecost.

CONTACT INFORMATION

www.jessikasatori.com
jessika@jessikasatori.com

206-463-7363
206-GOD-SEND

Made in the USA
Monee, IL
13 September 2026

62980997R00039